It's About Time!™

. . . All About the Months

Joanne Randolph

PowerKiDS press™
New York

Published in 2008 by The Rosen Publishing Group, Inc.
29 East 21st Street, New York, NY 10010

First Edition

Book Design: Kate Laczynski
Photo Researcher: Nicole Pristash

Photo Credits: Cover, p.1 © The Image Bank/Getty Images; pp. 4, 6, 8, 10, 12, 14, 16, 18, 20, 22, 24 © Shutterstock.com.

Library of Congress Cataloging-in-Publication Data

Randolph, Joanne.
All about the months / Joanne Randolph. — 1st ed.
p. cm. — (It's about time)
Includes index.
ISBN-13: 978-1-4042-3769-8 (library binding)
ISBN-10: 1-4042-3769-0 (library binding)
1. Months—Juvenile literature. 2. Calendar—Juvenile literature. I. Title.
CE13.R36 2008
529'.3—dc22

2006037191

Manufactured in the United States of America

Contents

There are 12 months in a year as time marches past.

We go from January to December, the first month to the last!

Four months last for 30 days.
Most months last 31.

Which month do you like the best? Which is the most fun?

I
LOVE
YOU

January is the first month.
It gives the year its start.

February is the next one.
We mark it with a heart.

March blows in, windy and cold.

Rainy April comes next,
as springtime takes hold.

May is the month when **flowers** grow tall.

June marks the end of school and fun for one and all.

July **fireworks** boom and light up the skies.

August brings cookouts and flashing **fireflies**.

September's a month when
the weather cools down.

We pick **pumpkins** in October
and dress up as a clown.

We give thanks for November
and eat lots of food.

Cold, snowy December brings
a holiday mood.

The months help us mark the time as the year passes.

We can look forward to our birthday or the start of classes.

Now you know more about the months than you did before.

What else do you want to know? Try to find out more!

Words to Know

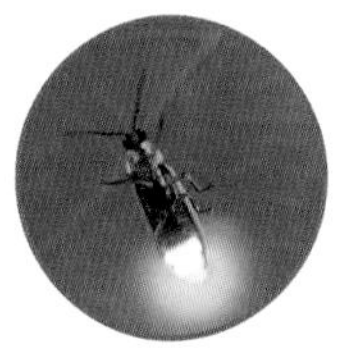

firefly

fireworks

flower

pumpkin

Index

Web Sites

Due to the changing nature of Internet links, PowerKids Press has developed an online list of Web sites related to the subject of this book. This site is updated regularly. Please use this link to access the list:
www.powerkidslinks.com/iat/months/